The Australian Guerrilla 5:

LURKING
DEATH

True Stories of Snipers in
Gallipoli, Sinai and Palestine

Ion Idriess

ETT IMPRINT
Exile Bay

This edition published by ETT Imprint, Exile Bay 2020

Also by Ion Idriess

Shoot to Kill
Sniping
Guerrilla Tactics
Trapping the Jap
The Scout

First published 1942 by Angus & Robertson
Facsimile edition published by Idriess Enterprises 1999
Electronic edition published by ETT Imprint 2020

ISBN   978-1-922473-25-7   (pback)
ISBN   978-1-922473-26-4   (ebook)

ETT IMPRINT
PO Box R1906
Royal Exchange NSW 1225
Australia

Photographs on p58 and p78 courtesy the Haydon Archive.

Designed by Tom Thompson

CONTENTS

Billy Sing, Sniper with the 5th Light Horse.

CHAPTER I
Abdul the Terrible

HE was death, hidden in a holocaust of life and death. Invisible, though many a nervous eye glanced up towards Dead Man's Ridge - some for the last time. They never found him. Strange, for Dead Man's Ridge was bare and sear while Australian infantrymen crouched staring at it from Pope's only a hundred yards away. New Zealand eyes searched it from Russell's Top. That an enemy could lurk there unseen was uncanny.

Yet day by day death spat from there.

Abdul the Sniper was the pride of the Turkish Army. They named his rifle "The Mother of Death". "Because," so declared the Ottoman Guard, "her breech gives birth to bullets which destroy the lives of men."

Though you were a thousand yards away you were face to face with him for his bullet bridged distance in a breath. His cold eyes flashed the deepest brown, heavy black eyebrows shielding the sight of a hawk; his swarthy face had the broad, hooked nose of a man-killer. His black moustache shaded a mouth with thin lips tight shut. Strong, lean brown hands rested caressingly upon his rifle, his trigger touch gentler than a lover's kiss.

Only two Australian enemy ever gazed directly upon Abdul, and lived: I - through a telescope, the other - along the sights of a rifle.

Many a man was to die first.

On this beautiful morning down in Monash Gully, a young Australian suddenly collapsed to the hard brown earth. That was all we knew the sniper by. The "smack" of his bullet, the spurt of rich red blood.

Far up on the Heights Abdul slid back his rifle-bolt, leisurely shut it again, the faint "click" music to his ears as the live bullet slid into the breech. Never had fox scooped out a fox-hole more craftily. In safety and comfort he lay there day by day gazing down at the great gully alive with scurrying victims. Roofing that fox-hole with a ground sheet daubed yellow and brown and green, part of the earth-invisible. His cap so dull and crinkled and stained was a clod of Turkish soil. His face and hands of living clay stained with the earth upon which he lay. Only his restless, wonderful eyes and the tiny, round, black muzzle of his rifle were natural.

Suddenly his eyes became tiger's eyes, his rifle-muzzle slowly rose, the veins tautened in his steady hands. One eye slowly closed, the other glowed terribly alive as his cheek pressed firmly low behind the rifle sights. By a hairbreadth the foresight rose to the level of that terrible eye, hovered there until the point of metal and eye were levelled with the head of a man far down in the gully below. A second's breathless pause - it meant the hush of eternity to one - then a spurt of vapour, a muffled report, and an Australian soldier sank to the ground.

Abdul the Terrible softly rested his rifle before him. Again he had taken the life of a man, it was the spice of life to him. He had fought in many a campaign, fought the

Greek and the Russ, the Bulgar and the Arab. He had been born a man-killer; he would die one.

He had never fought the Anzacs before, never even heard of them until that whirlwind landing several weeks ago. These men came to kill with a laugh, and the only way to stop them was to silence the laugh. Abdul was thrilled. He sensed that this campaign was to be the campaign of his life.

All along the line, fading into distance, sounded the rattling of rifle-fire, the stuttering of machine-guns. Down on Anzac Cove toiling men scattered to a screech, then Crash! and the Beach was sprayed with shrapnel.

"Beachy Bill" was sniping too.

Abdul the Terrible watched and listened appreciatively as a British cruiser roared into action; the Turkish Olive Grove thundered reply as the Anzac guns joined in. A roaring, howling, crashing, screaming artillery duel.

Among these big fellows were snipers too, giant snipers, sniping guns. But the might of the British Navy, the French, the British and Anzac troops could never silence the cunningly concealed Beachy Bill. We knew his whine coming across the hills far away. His sniping shells on Anzac Beach alone cost us more than a thousand casualties.

Turkish riflemen snipers took terrible toll of Anzac, especially from the Heights overlooking Monash Valley. They "bled" us day by day; their tally grew to grisly proportions. Expert riflemen, each fought a lone battle often in advance of his own ranks, actually at times, concealed behind us. No one, from private to general, was safe from the sniper.

On yet another day Abdul was gazing from his eyrie as a dreaming eagle might gaze down on to country far below. He thought of himself as an eagle, a terror of the Turkish mountains gazing down on the lambs clinging to the shelter of the farm lands. So were these Anzacs away below clinging to the shelter of the gully sides, clinging to their trenches, their saps, their barricades of sandbags. Like rabbits ceaselessly toiling up the great gully, carrying food and water and ammunition to their comrades on the Heights above. Their crouching little figures toiling ever upward to pause then swiftly run across the gully to shelter of sap or barricade. Then come toiling on until they must cross the gully again. Let one pause only a second if not under cover, let one incautiously show his head above sap or barricade, Abdul's eyes flashed, then his cheek sank to the rifle, one eye closed as the little black muzzle pointed straight down the gully. A spun of vapour, a muffled report - the eagle had pounced again.

General Bridges and several officers were walking up the gully, *enroute* to General Chauvel's headquarters. Bridges had endeared himself to the Anzac Army. The officers met Major Glasgow of the 1st Light Horse.

"Be careful of the next corner, sir," he warned. "I have lost five men there today - a sniper!"

Bridges nodded and walked on. Colonel White smiled. The general was noted for insisting that every man of every rank should take advantage of every possible cover, should never give a sniper a chance. But the general laughed at danger; he bore a charmed life. White uneasily thought he would never see the general take cover.

But Bridges, striding on, was thinking of the awful casualties caused by snipers, those devils who fired with such deadly accuracy from their fox-holes in the Heights. As he constantly warned his men against them, it was up to him to practise what he preached. They came to the dangerous corner and there, crouching behind a sandbag barrier, a group of Australian infantrymen called warningly to the officers.

"Better run for it, sir; it's very dangerous this morning."

Queer! throughout their walk they had each sensed some queer, strained feeling throughout the valley. It showed in the faces of the men toiling there, in the suppressed caution, in the "quietness". There was still the rattle of rifle-fire from the Heights, the stutter of machine-guns. Shells crashed above the Beach away behind them, while overhead and around was still the zip, plop, zip of bullets striking earth or sandbag. There was the usual "canary" whistle of bullets just overhead, the whine of shrapnel, the crash of high explosive. Red Cross men were attending wounded; sun-burned infantrymen squatted in their hillside dugouts delousing their trousers. And yet, this morning death seemed brooding over the valley.

To the surprise, of his, officers, General Bridges ran across the gully. They followed quickly. Several times they ran across, hugging the protection of the barriers. They arrived at the barrier below Steele's Post. The general paused to light a cigarette and here a crowd of wounded men earnestly assured him that the gully was terribly dangerous this morning. He must cross the gully again. They implored him to run.

Far up on Dead Man's Ridge Abdul the Terrible laid his cheek against his rifle. Steadily the muzzle levelled, pointing down.

The general ran - just those few yards to, shelter. He never reached it: As they carried him back to the shelter of the traverse he ordered huskily:

"Don't carry me back down the gully, Just leave me here. I don't want any of the stretcher-bearers hit."

But Colonel White stopped all traffic in the gully. The Turks would thus know that only wounded men would be moving. The rest was up to Allah.

Slowly they carried the dying general to the Beach. And the Turks did not fire a shot. He was bleeding to death, but he smiled up at Colonel Ryan and proudly whispered:

"Anyhow, I have commanded an Australian Division for nine months."

Billy Sing, in the trenches at Gallipoli 1915.

CHAPTER II
The Aussie Sniper

SOON afterwards there howled down upon us the Turkish offensive to "drive the Anzacs into the sea". With the roar of a bushfire the night burst into flame as brigade after brigade came charging, only to melt and die at the very edge of the Heights. To a shrieking tornado of "Allah il Allah" they came again and again. Night after night with the glint of their bayonets all red in the bomb bursts. Great men, these fierce-eyed giants of the Ottoman Guard. Very different fighting this to the crafty battling of the snipers.

Specialists like Abdul the Terrible the German-Turko Command thought far too valuable to be used with the attacking troops. The snipers were ordered to the flanks to kill, but not be killed.

After all was over the position was the same except that the flower of the Turkish Army was no more. The snipers crept back to their fox-holes. Unseen death again reigned over Shrapnel Gully, while up on the Heights the Anzac and the Turkish Army wrestled day and night.

To kill the snipers, snipers were chosen. Lieutenant Grace was ordered to organize an experienced band of riflemen against the snipers.

Each sharpshooter chosen had an observer beside him with a telescope. These sharpshooters, operating widely apart, crept out at night; each selected his own cover, or picked some commanding position in a trench. Thus they formed a long, rough line of crack shots just below the head of Monash Gully. Another line of them was posted throughout the valley. The tops of the enemy parapets, the numerous ravine heads, the scrub-covered crests were all before them. Each sharpshooter had to kill the enemy sniper directly before him, each observer had first to find that sniper. Thus, each pair of men concentrated only on a few square yards of ground away ahead towards the Turkish front, just where they thought a sniper was operating.

Minute after minute, hour after hour, day after day those hidden telescopes were searching those tiny areas of grounds. Searching for a bush that might not be a bush, for a fox-hole that might not be a fox-hole, searching for a spurt of vapour, for the slightest unguarded movement of elbow, face or hand, searching greenery that might be painted greenery, searching the parched earth that might be the stained face of a man. Those telescopes brought right up to the very eye of each observer the earth directly in front of him as if it had been an insect he was examining under the microscope. When the observer "spotted" a sniper he whispered to the sharp-shooter beside him. Then it was up to the sharp-shooter.

Thus, week after week, the duels of the snipers went on. In this fascinating, terrible game of man hunting man Australian sharpshooters were killed. But very soon there proved to be men among them who also were born to the game. They had never hunted men before. But now they were given the practice. Soon, very soon, they began to kill the snipers. Time and the game went on until eventually the enemy snipers were killed or driven elsewhere.

Abdul the Terrible was neither killed, nor found out, nor driven away. He was ordered away by the Turkish Command, one of the last to leave. For faraway out on the flank, somewhere opposite the Balkan Gun Pits, there had come a terrible Australian sniper. Day by day he killed man after' man and in the Turkish trenches too. Shelter meant nothing to this killer. A Turkish soldier need not necessarily be even manning a loophole, he would be killed just the same. The deadliness of this sniper soon got the nerves of the Turkish soldiery on edge along these particular lines of trenches. Abdul the Terrible was ordered to locate and kill this Australian sniper. Abdul had been decorated personally by the Sultan for his efficiency, and acts of sniping bravery. Now was his chance to prove that the Sultan's choice was not misplaced. Abdul was ordered to go and get his man.

All these things we knew later, from the gossip of prisoners, and from diaries which some among the Turkish soldiery were fond of keeping.

With all his experience and cunning it took Abdul the Terrible a long time indeed to locate his man. He was in Chatham's Post: trooper Billy Sing, of the 5th Light Horse, probably the most dangerous sniper in any army throughout the war. A born sniper, active service had put the finishing polish on him. Used to Australian distances, to the faintest movements of the bush, a student of camouflage in Australian animal and bird and reptile, he handled a rifle as confidently as another man would handle his pipe. Possessed of extraordinary patience, he had nerves of steel and eyes keen as Abdul the Terrible's.

I "spotted" for Sing one day. Comfortable within his cunningly camouflaged possy he watched the Turkish trenches as a cat watches a mouse. If nothing much was doing he would relax and the spotter would take over the job. Behind the craftily concealed loophole, he'd train the telescope on to No Man's Land and the indistinct array of trenches that shielded the Turkish soldiery. It was very much like a cat watching broken walls in which there were many mouse-holes. Behind every such hole, many of them invisible except to a telescope, was a cautious mouse. Should the telescope spy the faintest sign of movement behind a hole, Sing would become instantly alert; glance through the telescope, then stealthily train his rifle on the loophole.

How many enemy this particular sniper shot will never be known, but in three months his tally was one hundred and fifty. Each was authentic, seen by an observer through a telescope. No hit was counted unless thus definitely witnessed even though the sniper fired often when alone, and knew that he had registered a hit. This man who officially shot one hundred and fifty men was the sniper whom Abdul the Terrible must kill.

New Zealand sniper and his spotter at Gallipoli.

CHAPTER III
The Duel

ABDUL the Terrible sought to locate Sing. He inquired for him throughout the Turkish Echelon Trenches, he walked through the Turkish Despan Works seeking information of officers, non-commissioned officers and men. Thoughtfully he knelt by the grey-coated infantry in the Balkan Gun Pits, staring out towards the Australian trenches.

All that the men could tell him was that death came from the trenches somewhere in front where the brown men were who wore feathers in their hats. But just where it came from they did not know. They did know that now every man feared to face his loophole.

The officers localized the position better. They thought the deadly sniper was somewhere in those trenches running downhill to the sea which the Australians called Chatham's Post. But they all ended with an expressive gesture and:

"He is out there - somewhere. Find him."

When a man was shot through the head Abdul would hurry to the spot and examine him. If the bullet had taken him right between the eyes-that would be frontal. If in the left temple-then the shot must have been fired from away to the left; if the wound was in the right temple or the right corner of the forehead then the bullet must have come from the right. Sometimes the bullet had struck right down through the top of the head. Abdul would reconstruct the shot, it might mean that the victim in a moment of carelessness had dropped something and while in the act of looking down at it or bending to pick it up the bullet had come through the loophole. He would look for marks in the trench that might tell whether the soldier had been standing, lounging or stooping. He would try to visualize the precise angle at which the bullet had struck; whether from a straight level, or below or above the level of the loophole.

With those details, his experience gave him a surprisingly accurate idea as to the direction, angle, and distance from which that particular bullet probably had come. Not then, but an hour later, with his sniper cunning he would peer from the loophole. Not as a soldier might peer but as a sniper would peer. Every sense alert, with his roving dark eyes staring towards where that bullet must have come from.

Perhaps he would immediately realize that the bullet could not have come from the short line of enemy trench visible through that particular loophole. He could thus eliminate that particular portion of trench. Then his glance would slowly sweep across No Man's Land, seeking their objective more cannily by far than a searchlight would seek a hidden plane in the sky.

Thus Abdul investigated the death of man after man, in trench after trench. With one detail gleaned from the death of this man and another from the death of that, he gradually pieced all the evidence together. And each bit of evidence was true evidence, for Abdul was rarely called to view any man not hit by the sniper. It is uncanny how very soon and consistently men in trench warfare learn to tell whether a particular sniper or just an ordinary soldier is firing at them. Thus every victim left his voice-less story, and all gradually merged to point in the one direction.

As time went on Abdul eliminated various portions of trenches; was not sure of other portions; was becoming increasingly sure of one portion. His countrymen's trenches were dug in different positions at differing levels. The enemy sniper must necessarily be hidden in a position which commanded these various levels and differing ranges, else his bullets could not enter these positions. Surely and yet more surely that definite angle of direction was taking shape in Adbul's mind, like a straight, invisible wire pointing more and more towards one spot nearly on top of a trench across on Chatham's Post.

Abdul at last was satisfied. Now he could choose his own position, a position overlooking the low ground between, that would directly command that spot opposite.

This meant a duel to the death. At night Abdul began to dig his cunning fox-hole, leaving it before dawn exactly as it had been apparently. Even to a telescope from the opposing trenches every bush, every pebble, every coarse blade of grass appeared exactly as it had been.

He finished the fox-hole. It was invisible. He now climbed into it before each dawn and lay there all day, staring across at Chatham's Post. The long brown line of irregular parapets, the camouflaged loopholes behind which men crouched. He did not fire even though occasionally a tempting target offered itself. He sought now the life of just one man.

For a long time he could not find him. He could see numerous loopholes and closely studied each as again and again a man fired. But Sing's possy was camouflaged so naturally that even did anyone see a shot come from it how was it to be known that there lay the sniper? One evening, however, Abdul climbed down into his own trenches and walked noiselessly along to the officer in charge.

"I have found him," he reported gloatingly. "I will kill him - tomorrow."

"Good!" replied the officer. "It is the will of Allah."

Next morning the Australian sniper did not climb up into his possy as usual. He could please himself; his job was specialized and entailed a constant nerve strain, so he was free of the ordinary routine duties of the soldier. Most of us in the firing-possies were amusing ourselves by smashing the Turkish loopholes. These almost invisible apertures were neatly framed by four bricks upon which the sandbags rested. We were sniping those bricks and as the bricks broke so the sandbags fell down, blocking the loophole and driving Johnny Turk crazy with rage. A couple of hours after Stand-to, on some mornings we'd shoot away the bricks so that for hundreds of yards there wasn't left one serviceable enemy loophole. Old Jacko must build them up again at night.

This morning the Turkish bullets came viciously back

at us, growing less and less as we smashed loophole after loophole. It was great fun, but only to the highly trained soldier. We knew how to guard ourselves in thus playing with death. During those times I sometimes thought how awful it would be for untrained troops to face a trained enemy.

One of the boys climbed up into Billy Sing's possy. He had no right to be there; for the tiniest miscalculation on his part might betray that possy to a watchful sniper. But you know how these things happen. Sing's possy commanded more of the enemy's trenches than ours did, hence there was better shooting up there. The Aussie climbed up, cautiously opened the peephole of the loophole a little way and slowly poked his rifle-muzzle through. Then, as he saw a target - fired.

What possessed Abdul the Terrible to hold his fire, no man will ever know. Instinct, inborn knowledge, the intuition of the master craftsman must have told him that it was not the real sniper firing there.

Presently, Billy Sing appeared. He sat down on the firestep, yawning. He was not so bright today. The trespassing Aussie climbed down, the observer climbed up with his telescope. With a grunt Sing followed, stretching out beside him, half yawning as he touched his rifle.

"Heavens!" exclaimed the observer. "Quick! Look here! Don't open the loophole whatever you do."

The sniper was instantly wide awake. He leaned across to the telescope. Stared motionless.

What he saw was a just a face, a man's face, brown and stained, a big nose, a black moustache, two big, dark staring eyes - and the little round circle of the rifle-muzzle. Thus the Australian sniper stared into Abdul the Terrible's eyes.

"Spot him?" whispered the observer.

"Yes," answered Sing slowly .

"Only by the merest chance," thought the observer, "one chance in ten thousand, I put the telescope right on his face and casually glanced. Right on his face! There are all the trenches and the Ridge, the Balkan Gun Pits and the Bird Trenches and the Echelon, and the snipers' possies out in front and all the country in between, and yet I put the telescope right on a hidden man's face!"

The sniper glanced around to see that his possy was dark, for not the slightest gleam of sunlit space must betray the tiniest opening of the loophole. The observer stared through the telescope. The sniper, with his finger, slid back the loophole cover hardly an inch, then cautiously poked his rifle-muzzle through, the rifle that had taken the lives of one hundred and fifty men, of far more men than that.

"Careful!" murmured the observer. "He's got the eyes of an eagle and - he's staring straight here."

"It's me or him!" grunted Sing.

But had Abdul fired, even had his bullet come through that tiny slit, it wouldn't have hit the sniper, for the born sniper knows the crouch that means the fractional difference between life and death. Only when the sniper actually had his eye aligned with his rifle sights then-

That was what Abdul was waiting for. His big eyes staring, his rifle-muzzle slowly rising up .... But Abdul did not know that the Australian sniper had seen him.

Gently the peephole widened, then stopped close around the rifle. Abdul waited with finger on trigger, just awaiting that loophole to open the least fraction more. And - a bullet took him between the eyes.

Australian snipers at Gallipoli.
Billy Sing, taking a rest at Gallipoli.

CHAPTER IV
The Raid

ASTRIDE his camel, Kara Ismet mused on the dawn while riding under desert stars. Ahead of him was the long line of camelmen. His dark eyes glanced to right, then left. To the right another ghostly line that seemed pan of the desert and the night. Then to the left, a long dark smudge like a moon-shadowed cloud crawling upon the sands. Infantry these, pressing forward to attack Oghratina. Kara Ismet thought gladly that at least he was not marching through the sand. An all-night march might affect his aim. At dawn his officer insisted his aim must be steady, particularly quick and deadly. Kara Ismet frowned, for it might be he who would be shot. This affair of raids with point-blank shooting then the sudden rush was not a sniper's work.

Kara Ismet the sniper was a brave man who, having dealt death so often, knew that death comes but once to any man and - he did not want it to come to him.

"Allah!" he mused. "Twill be the will of Allah," and his cyes reflected a star-beam as he glanced at the sky.

A broken-nosed Anatolian was this Kara Ismet, a Russian rifle-butt had flattened his nose like the smack from a horse's hoof. A nasty scar ran right across one cheek. Souvenir from a

Greek bayonet-thrust this scar; he had shivered at the sight of steel ever since. His comrades called him "The Hyena", for when he laughed that scar twisted his face into a snarl. A swarthy face with black moustache and eyes quick and bright and keen. A good soldier was Kara Ismet, and a wonderful sniper.

The raiders pressed on in a faintly murmuring silence that was but the "Squelch! Squelch!" from the camels' pads and the soft sigh of the infantry's boots in the sand. Determined and deadly efficient these picked troops of the Sultan adventuring far through the desert in what, unfortunately for us, was destined to be one of the most successful raids of the war.

The Turks were supposed to be far away in Palestine. Actually these men were well into Egypt and within twenty miles of the Suez Canal.

It was some months after the Evacuation of Gallipoli. The British Army was spread out in strong-posts, mainly along the length of the Suez Canal, while the Australian and New Zealand mounted men were now riding down to reinforce them. On this very night, when Kara Ismet rode so thoughtfully with the raiders, my old regiment was riding thirty odd miles away, towards them. If only the British Command had given us the order to march but eight hours earlier!

The 5th Brigade of Yeomanry were stationed out in the desert miles away from the Canal. They had formed outposts of isolated squadrons some six miles apart among the scattered palm oases. It was these lonely Yeomanry posts, in particular, that the night raiders were now marching to attack. Cheery English boys this 5th Yeomanry Brigade, but alas, untrained in desert warfare.

The stars twinkled down upon a shadowed sea of loneliness far removed from the world of men.

Kara Ismet's pulse quickened as he noted the camels ahead beginning to turn just a little to the right, while the camels to the right were imperceptibly fading among the small desert bushes. And now that shadowy line of voiceless infantry to the left was melting into the desert. So the columns were diverging, each to their separate attack. It would not be long now.

Allah blessed the raiders. Instead of bright dawn, an almost impenetrable fog shrouded the desert. Fog of the desert! And that it should come on a dawn such as this!

The raiders dismounted. Shrouded forms remained among the kneeling camels while the fighting men crept forward, treading noiselessly behind their Bedouin guides. Friends peered towards one another's faces and voicelessly smiled, a deathly excitement in their eyes. Miles away across the desert their fellow raiders were creeping forward similarly, surrounding three other such posts, their bayoneted rifles gripped at the ready to thrust into the outposts. But not an outpost! ... Not a sound! ... Not a rifle-shot. Breathlessly they crept forward to instinctively halt. The Bedouins were pointing for the officers, the Bedouins' eyes were gleaming as they whispered into the ears of the officers, their fingers like nervous claws around their dagger hilts.

Like a crouching panther the whole line listened, a long, long time. They could hear the murmur of voices. Then laughter, as out there in the mists a Yeomanry lad cracked an early morning joke. The raiders trained machine-guns towards the sound; every rifle was levelled; a volley startled the misty air. To the chattering of machine-guns the raiders rushed forward, and the fog lifted, as if by magic, showing a group of startled men with open-mouthed, amazed faces. Among horses and saddles and tents hundreds of sleeping men burst from their blankets. A wild command, snatching for rifles, stutter of machine-guns and the raiders dropped to cover, firing rapidly.

Each attack was complete and an utter surprise. But those of the Yeomanry left alive after the first burst of fire scooped holes in the sand and each man fought to the last. Sometimes the fog rolled away and the rifle-fire burst out snappily, then the fog rolled back and mercifully blotted out everything. Kara Ismet, his eyes roving to pierce the fog, could only see indistinct targets that swiftly fired back. He could not pick out his particular prey, the officers and machine-gunners. The Turks, brave men, but hating to meet the bayonet, kept firing and picking off man after man until the Yeomanry fire was only a splutter - their ammunition was finished. Then with a wild yell of Allah! Allah! the raiders rushed in and finished the fight,

Thus at Oghratina, and thus at Bir-el-Katia. But at Bir-el-Duiedar they were beaten off. Duiedar Oasis was held by the Scotties and Yeomanry, the Scotties' foxy rushed out and barked furiously. The poor little foxy was immediately clubbed by a Bedouin but the alarm saved them.

It was my old regiment that galloped across the desert to the rescue. The Yeomanry lost six hundred men in that raid. The little holes that each man had dug in the ground, the little heap of empty cartridges, the dead horses and smashed camps told of a bitterly contested though successful raid.

The raiders vanished, and Kara Ismet with them.

How he hated raids - and above all the steel! He was a sniper. He blessed that fog. But for the fog they would have had to finish this raid with the steel alone.

Kara Ismet shivered. It is one thing to give a man the steel but quite another to take it.

A Turkish Sniper wearing camoflage.

CHAPTER V
An Army Marches

KARA ISMET on outpost duty munched a handful of dates. He was perched like a crow upon a giant sandhill from where his eyes lazily roamed over the desert. Distances stretched far over innumerable sandhill crests under the blazing sun. In brilliant distance there stood, like a tiny deer, a Bedouin scout with his camel. Golden sandhills stood everywhere crested above valleys of sand pitiless to life. In drab patches here and there were dull-grey desert shrubs, their roots all mounded by windblown sand.

Kara glanced down into a great valley. A big, dark grey patch, stretching back for miles, was toiling slowly along. A wonderful little army this; it had already marched a hundred miles across desert from its base in Palestine. Toiled over those terrible wastes, and dragged its guns with it. Even the British aeroplanes that had droned overhead had not seen it, so cleverly was the march organized. Kara wondered at that organization while watching the men toiling at the guns. Those sullen, heavy wheels were not sinking into sand, they were creaking over planks.

Sweating men were carrying planks and laying them in front of the gun wheels. As the wheels rolled over them other men picked up the heavy planks, shouldered them, and marched doggedly past the guns to again lay the planks down in front. Farther along the valley men were digging shallow wheel tracks and packing them with desert bush to make foundation for the wheels of the guns. What a surprise it would be for the British, to find that an army with guns had crossed the desert! They made Kara confident of the future, those guns, manned by expert German and Austrian crews.

It suddenly crossed Kara's mind that no German was toiling down there with the heavy planks, nor hauling on the guns. The Germans were giving orders; they always did. The Turks were the toilers and the infantry fighters, experts with the rifle, but indifferent mechanics. It was right that every man should be in his place, at his own task.

Kara thought of the German machine-gun crews, of the German aviators. Everything that was mechanical, was German. German telephone corps and signal corps and - yes, away back in Palestine all the motor transport was German. Kara idly thought that all this mechanism, the aeroplanes, the artillery, the machine-guns, the signals and communications was the brains and force of the army. The dogged Turkish infantrymen went into battle directed and supported by and - yes! in front of German mechanism. Kara stared hard. As a sniper he must always think out his own plans, his own individual advances and retirements; it now occurred to him that these few Germans actually controlled the Turkish Army; that their Intelligences ordered the advance, their machine-guns and artillery covered it. But - the Turkish Army could not turn back unless the German machine-gunners and artillery were willing!

Kara dismissed such foolish thoughts. This was an army marching to victory. There would be no retreat.

He stood up and walked solidly on, high up among the sand-ridges. He was one of the eagles, there were numbers of them far to left and right, eagle-eyed snipers whose job was to detect and shoot enemy patrols - if any.

There weren't any. The plans of the desert army had been organized too well. The British still thought them far away in Palestine.

Night and day the army toiled stolidly forward, living on but a sip of bad water per day, and a handful of dates or pressed apricots. Veterans of Gallipoli these, with picked battalions of fighting Anatolians; all buoyed up with a victorious morale and a dream-land of the future ahead.

They had won at Gallipoli, beaten off the crack battalions of Britain and France, had beaten too the Russian hordes. And now had just come the grand news of their smashing of the British Army in Mesopotamia.

And but a few months before, there was their vic-torious desert raid on this new front. All they had to do now was to smash the British Army at Romani, then press on to the Suez Canal and Egypt.

The rape of Cairo and Alexandria had been promised them, together with the fabulous spoils of the rich Egyptian delta. They were marching to a paradise of loot. They had trudged through hell in that desert march, uncomplaining. Paradise awaited them at the end.

But the gates of earthly Paradise take entering. So thought Kara Ismet, weeks later, under the stars lying on the chilly sand by Oghratina. Oghratina! where but a few months before he had seen the untrained British Yeomanry fight to their last cartridge, their last man. But it was not an isolated post that lay before them now. The Anzac Army had spied out the Turk. Around them spread the great Karia Oasis system, stretching for miles in scattered clumps of oases, each with its hundreds of date palms, its priceless well of water. Some miles behind these cool, green oases there loomed up the great sandhills of Romani. Those grim ramparts of sand towered up to overlook the oases and the desert stretching far away.

The Turkish Army now was within twenty miles of the Canal, the toil-stained soldiers could feel Egypt within their grasp. Only Anzac cavalry lay between them and their dreams. Cavalry! Had they not massacred British cavalry at Oghratina and Katia?

Kara Ismet, staring into the desert, warmed at the thought of meeting these Anzacs again. He had shot them at Gallipoli; he would shoot them now. These men of the felt hats would learn that desert fighting was very different to fighting with their great warships behind them. Idly he wondered what manner of men they were on horses.

He soon knew.

Next morning the line of Turkish snipers was ordered out to the front, each to pick his own foxhole and shoot any oncoming patrols, thus warning and protecting the infantry now feverishly digging in a line of redoubts stretching for miles across the desert. For the clever commander, though not dreaming of defeat, was determined that, if necessary, his army would have entrenched defences to fall back upon. That thought, followed by action, was destined to save half his army.

Kara Ismet selected his bush, it was just below the skyline

of a sand-ridge. A pile of sand buttressed the scraggy roots. Kara would scratch out a hole here and shelter under the bush, it would afford him both cover and protection. But he was dissatisfied with the view in front. It showed numerous large desert bushes with, in the distance, the huge mass of date palms in the Katia Oasis. But the country directly ahead of him was not clear, it was cluttered with bushes. However, he did not think there would be much doing. He laid down his rifle and knelt, lazily scooping away the sand. It was hot.

The intuition of the born sniper set him back upon his heels, his ears seeking every sound. He wheeled around, fearful amazement upon his face.

Among the bushes came riding four scattered horsemen who rode to act as one. Never in its immemorial history had the desert seen such horsemen as these. Strong and lithe as if part of the very animals they rode, they were burned black, almost as Bedouins. One brown hand clutched reins that in an instant could wheel the horse into a gallop; one bare brown arm held a rifle that could be swung into action in a second. Each was clad only in flannel with a bandoleer of cartridges, a broad bayonet belt around worn riding-trousers, and an old felt hat shielding a hard brown face with eyes keen as snipers.

"Anzacs!" thought Kara and whipped up his rifle.

But the horses plunged forward and were galloping through the bushes. In an instant they disappeared. Kara crouched breathlessly.

"Crack! Crack!" Quick rifle-shots and again "Crack! Crack! Whi-zzzz!" a bullet smacked the bush by Kara. He leaped and, crouching low, ran to tumble down behind another bush with a galloping thud in his ears as he rolled and rolled to another bush with a horse's belly leaping over him and a rifle firing straight down at him.

In a second the horsemen had gone leaving Kara squirming farther among the bushes. He listened, his heart thumping. They had gone: he could not even hear the sound of their thudding hooves. He listened and listened. Miles away, sounded "plip-plop, plip-plop" of snipers' rifles. So other horsemen were spying far away down the line too!

Cavalry! What manner of cavalry were these?

Never a chance of a shot at them! And how they could gallop and shoot!

Kara crouched there not yet venturing to peer above the bushes. He might be sniped at! He peered uneasily around, for no sniper likes being sniped at. Those four horsemen had appeared and in a second had galloped straight towards the sniping line thus drawing the sniper's fire while galloping swiftly on, searching the country behind to see what the infantry were doing. At the gallop they had wheeled and were galloping back again but - in another direction. And now were galloping to headquarters with the information they had so quickly gained.

Kara wondered whether he would have to learn sniping all over again.

CHAPTER VI
The Desert Riders

THUS began a hell of heat and thirst for Kara Ismet and for the remnant of the Turkish Army fated to survive. That army was spreading its vanguard across the desert, shielding the battalions piling up behind. But at any moment by day or night, in troops, then squadrons, then regiments, the Anzacs would suddenly appear at a mad gallop, leap from their horses and, pour in a furious fire. When the panting Turks came close they would gallop away only to dismount and open fire again until the Turks plodding after them would sink to the sands and gasp.

And snipers!

At the first crack of a sniper's rifle these men were galloping the bushes as if hunting a wild dog. No fox-hole, no hide-out however cunningly dug deep down under the bushes was safe from those plunging hooves, those searching bayonets. It was only while the sniper clung dose to his own infantry that he might continue firing when he saw targets, and survive. And the more the snipers were hunted the more furiously the officers commanded their snipers to keep well out in front and harry the horsemen's flanks as the grey wolf harries the deer.

Sullenly, one morning Kara had dug himself a naturally concealed fox-hole-well out in front. No man works so cunningly as when he knows his life depends upon it. The day wore on. For miles among the desert bushes, the palm oases, the sand dunes, came isolated crackling of rifles and snap of mountain guns that told so well the Anzacs were harassing the infantry. But all was quiet where Kara lay. He wondered when the battalions behind him would come moving on, when the army would have finished its feverish digging of redoubts preparatory to the grand advance on Romani.

Though Kara lay so quietly, his ears were listening to every desert, sound, his eyes searching the brown sand patches interspersed with desert bushes, his rifle before him ready. He would dearly like a shot, a clear shot, at these wild horsemen. It hurt his pride that he, during days past, with enemy all around him, had not yet killed a man. He stared, suddenly gripped his rifle. Bobbing up over the deeply broken ground and big-tufted hillocks appeared three felt hats, then numerous other brown faces to left and right, then rifles and the ears and noses of on-coming horses, the quick, soft pounding now of their hooves. A squadron of Light Horse led by a captain. Kara Ismet could have shot the captain with delightful ease had it not meant death for him too.

He lay as breathlessly as ever fox crouched to burrow, then almost shrieked. For only yards behind him he heard a whispering squelch that could come only from the boots of an advancing Turkish battalion. Horses' legs were striding past him, swishing among the bushes. His nerves were paralysed as the squadron rode right on into the infantry. Sudden suspense, a hundred hooves stopped short, then a wild yell and a hundred horses plunged forward to screams from Turks as they snatched at the reins only to be flung aside or clubbed as the plunging horsemen wheeled and with mad yells of laughter were galloping away.

The Turkish infantry stared after them, their limbs trembling, their faces comical in strained amazement. To a cursing howl from the officers they lifted their rifles and fired - too late.

Several days later precisely the same incident happened again, even more to the discomfort of Kara Ismet. Again he lay hidden out ahead of the Turkish screen, again that screen advanced and a squadron of Light Horse, a 7th Light Horse Squadron under Major Richardson, rode right into the Turkish screen.

Again that mad yell with the Turkish infantry snatching up at the bridles as the horsemen wheeled and were away. The Turks raised their rifles and blazed hysterically but did not hit a man, the Turkish infantryman when surprised always fired wildly. This time the Australians wheeled around when only three hundred yards distant, flung themselves to the ground and opened a quick, steady fire even while their horses were being galloped away. Kara Ismet found himself in the advance compelled to run forward under a hail of lead. Thus he threw himself to earth amongst the panting infantrymen and now bullets kept his head to the ground, spurted sand into his eyes, machine-gun bullets ripped swathes through the bushes. This was not sniping.

Again an officer's sharp order and the Turks leaped up and raced with bowed heads to drop panting again to the bullet-swept sand. Again they charged forward and this time, half-blinded with sweat, panting from thirst-blanched lips as their shaking hands worked their rifle-bolts. They knew this was the last time, the next time meant the charge and the steel.

Kara Ismet felt almost glad. This running over the burning sands under a deadly fire would burst his panting heart. Far better to get it over.

But while lying there panting for breath Anzac machine-guns kept their heads to the sand. Then horses were galloped up and in a trice those wild horsemen were mounted and away.

The staggered Turks raised unbelieving faces, stared out among the bushes, then at each other.

"Thanks be to Allah!" murmured Kara Ismet, and a bullet whipped the cap from his head. An Anatolian crouching near by looked amazed then gently his head sank to the sand.

Bullets whizzed amongst them, whipped the sand and bushes. Those cursed Australians had dismounted again; were fighting again; it had all to be done over again.

Sharp calls from the officers, again the infantry rose and the line ran forward, ran for dear life with bullets screeching past their belts. Again they fell flat to the burning sand and bullets kept them there while pant¬ing breaths and trembling hands helped speed their bullets wildly.

Again, and yet again, the Australian squadron fought with the same tactics until the Turks by mid-day could barely stagger forward. Then, to add insult to misery, these cursed infidels calmly boiled their quart-pots, ate their bully-beef and biscuits, while their machine-guns kept down the heads of the exhausted Turks. Officers and men just lay there under what cover there was, utterly exhausted, only praying they be allowed to rest a long, long time. Suddenly, in a whirlwind rush the Felt Hats were on their horses and away again only to open fire from a few hundred yards distant.

To a hoarse, croaking shout from an officer Kara Ismet wearily arose again. His stomach seemed afire from want of water, his lips were burning as he stumbled forward with the others. How could he obey that cursing officer, how could he possibly pick off those Anzac officers, those machine-gunners, when his own limbs were trembling from exhaustion, his eyes stinging with sweat, his body tortured with thirst. As the bullets whistled past him he mentally cursed this crazy officer who belied the fact that a sniper is a killer who kills only when body and nerves and mind are steady.

Thus all that day the squadron of the Seventh fought, as for miles along the great oases system other squadrons and regiments were fighting, as tiny sections and outposts and troops were fighting. This was merely preparatory work, keeping in touch with the Turkish line while endeavouring to lure their main army to the grim Romani sand hills miles away behind, them.

As the days went by, the main Turkish Army, their defences behind them now completed, began steadily advancing in an ever extended line, pushing on stronger and stronger.

To the Turkish infantry and Kara Ismet their hardships at night were now as bad as by day except that, thank Allah, there was not the same terrible thirst and heat. The Turkish line pushed on night by night, feeling their way across the starlit desert, their ears and eyes and nerves tingling for the expected thunder of rifle-fire at their belts. Their feet were almost noiseless in the sand. To Kara Ismet every big clump of bushes, the broken clumps in starlit silhouette, every group of palms developed the shape of those wild horsemen. His nerves tingled for the line of flame, the rifle volley. But not a sound, not a movement. Stealthily the shadow army moved on. Suddenly a breathless shock thrilled nerves into hysterical inaction, then a rush of hooves rapidly galloping fainter. Great relief: the Turks had walked right on top of a listening post.

Thus, in scattered posts for miles along the line, Australian horsemen had been sitting motionless, listening, counting the ghostly shadows creeping towards them, then galloping away from under their very rifle-muzzles. All along the desert line it was the same. No matter how cautiously the Turkish Army, or part of it moved, or where it moved, there were always these hidden Australians to detect it, then gallop back and hide and wait again. Thus, their headquarters miles back with their main army constantly knew the progress of the Turkish Army. Many a night march by many an army had those stars witnessed, but never a march like this. An army of desert men ceaselessly shadowed by phantom horsemen.

Kara Ismet one night, peering ahead while advancing with the line, suddenly wheeled with a half-choked shriek and threw himself flat as a horse galloped over him. Thunder of hooves, and they were away. Chatter of teeth. Then in relief the officer's furious question: Why had he not shot that patrol?

They had galloped right over him and he, a sniper, had not fired a shot!

Not a man in the line had fired a shot. With nerves strained and rifles front they had been creeping over the sands. Suddenly, these mad horsemen had plunged at their very backs. Their *backs!*

It happened so on other nights. The tiny Light Horse patrols, their listening posts and Cossack posts would be motionless amongst the bushes, even on the bare sand. Listening - listening. Not a sound. Suddenly shadows to right, shadows passing to the left. Too late to move. They waited, still as the very sands. The line of Turkish infantry advancing by their left, the line advancing by their right, noiselessly passed by. Quietly then the horsemen would mount and ride behind the infantry, following them, seeking a break in the line. Perhaps they could not find one. Sooner or later a nervy Turkish soldier would glance behind and immediately the horsemen were over their horses' necks plunging through the line at full gallop. They had vanished in seconds, leaving the Turkish soldiery half prostrated and in alarm.

It is nervy enough at night advancing expecting to walk into hidden rifle-fire at any second, but a hideous climax when horsemen suddenly plunge at your back.

It sometimes happened that an Australian patrol, when unknowingly trapped by the Turks, would ride behind the Turkish line until they found a break, then quietly ride through the break and away without the Turks ever knowing they had been advancing in company. Throughout that stirring time every minute of the days and nights was adventure for the mounted men. Almost unbelievable how few of these adventurous patrols and outposts were killed. Though several posts were surrounded and killed to a man, still hundreds of patrols by day and night secured their vital information then got safely away from under the very noses of the Turk.

Presently, the Turkish line was firmly established for miles across the oasis desert. Then the Turkish Army concentrated. The Big Push on Romani was coming.

Kara Ismet felt more satisfied now. Immovable infantry was behind the long, thin line of sniping nests. The snipers again were killing men. But they were being killed too, and a sniper hates being killed.

Sniper Billy Sing with the 5th Light Horse in the Desert campaign.

CHAPTER VII
Romani

BRILLIANT stars and a quarter-moon bathed the desert in ghostly light. In unearthly beauty the sand-dunes of Romani frowned upon the desert plain. The steep faces of these moon-bathed sandhills, apparently liquid gold, were really flowing sand into which a man sank to his knees. From the towering crests above, gentle slopes led back to still, black clouds that were waiting Australians, New Zealanders, and Yeomanry. High up, lying invisibly upon the silhouette of fantastic crests was a lamentably thin, miles long line of Anzacs. With rifles beside them, machine-guns set, they started out over the desert, never talking above a whisper. Grim-faced men these, as akin to the desert as Arabs but incomparably more formidable. Separating each great sand-ridge from its neighbours was a shadow, black as the pit. These were really the mouths of valleys through the sand, gateways leading back through the ramparts. Machine-gunners and riflemen waited down there too, for through these black open spaces an enemy could certainly attempt to rush. These ramparts of Romani were now all that barred the

Turkish Army from the Canal and Egypt.

The dreams of men are often built on shifting sands. And now the dreams of an army depended upon the sandhills of Romani.

Throughout a thousand thousand years, and over the armies of many campaigns, the sandhills of Romani have frowned down upon the great Katia Oases system and the priceless water there. Egyptian spearmen, Roman centurions, Arab irregulars, Napoleon's army - a thousand armies have marched and fought and died there. And this night witnessed a small army of the youngest nation on earth waiting to fight upon this oldest battleground on earth.

Far out in front of the Anzac line, Cossack posts and listening posts stood silently by their horses. Far apart were these tiny posts, waiting for the long lines of advancing Turks to creep upon them. Down there the desert was fairly hard, covered with tiny hillocks and bushes amongst which a silent enemy could advance unseen.

The far-flung Turkish lines were advancing in force. Kara Ismet stepped silently forward, he could see the few grey-clad comrades to his right and left. But courage whispered that those ghostly forms stretched away for miles to right and left, with supporting battalions following close behind.

Death hovered over the watchers on the Heights; advanced with the ghosts upon the plains. Brilliant stars from a glorious sky made sand and bushes and shadows a fantastic medley through which, far ahead, visions flitted. Kara Ismet's anxious eyes stared ahead and to right and

and left to rest distantly on the black shadowed bastions of Romani. They at least were immovable. Yet, even as he gazed, they seemed to move.

Silently, doggedly, their nerves all tensed, expecting action from phantom horsemen; the invaders marched on with the light of fanaticism kindling in their eyes. Overwhelm Romani; then - loot Egypt!

Suddenly - a rifle-shot. Two more! Thousands of men halted breathlessly. But no roar of rifle-fire broke the desert silence. It must have been some Australian outpost, overwhelmed in one swift rush.

The Turks marched on. Midnight came. The long line halted. Their creeping screen brought back whispering reports that the deep gullies leading into the sandhills were heavily held by Anzacs. The moon softly went to bed, starlight alone played on the desert. In silence the miles-long line of Turks lay on the ground. Another breathless hour went by. Then Kara Ismet shivered, for the faintest sighing like the breaths of many men told of thousands of boots coming. The main body had arrived. They were granted but short breathing time. Kara Ismet sensed rather than heard the whispered order. With tingling fingers he fixed his bayonet.

Then orders were shouted and thousand of Turks were racing forward screaming "Allah! Allah! Finish Australia! Finish Australia!"

From the black crests came a crackling rifle-fire roaring over miles. The Turks lay flat, firing at the flashes under a rain of bullets. They crept closer until Kara Ismet was screaming forward in the wild charge upon Mount Meredith. They plunged up to flounder, leap and jump, climbing and yelling, with men falling between their legs. In a hail of fire they almost gained the crest. Then Kara Ismet found himself rolling down, clinging to his rifle while choking in the terrible sand. Gasping men and dead men were rolling with him; clawing hands scratched his face. They melted back into the darkness to fling themselves in exhausted despair on to the harder, cold sand.

The far-flung battle-line was a roar of sound, beautiful with miles of spitting flame, whining with bullets, howling with the Turkish battle-cry. But everywhere the thin line of Anzacs held their ground. Again Kara Ismet was screaming forward in another frantic charge up the cool, golden face of Mount Meredith. This time they gasped to the very crest. Then starlight gleamed upon· steel as phantom men arose. With a terrible shout they leaped forward and steel was plunging into Turkish chests. Kara Ismet's bayonet was clashed aside and he slipped, rolling, rolling, rolling down that wall of sand.

The charge was beaten back. But reinforcements rushed to the Turks and they came gasping up the sandhill again. Again their charge was beaten off.

Several hours later irresistible waves of them swarmed up the narrow valleys and took Mount Meredith in the flank. To exultant Turkish yells the shadowy Anzacs were seen running back down the hill. The Turks howled after them as they vanished into the blackness of their horses. They were upon them and away, clubbing down with rifle-butts as the startled Turks bayoneted the horses. Kara Ismet heard an Australian officer shout and four dismounted men leaped towards him, two men each grasped a stirrup-iron and the five were away. Dodging rifles, the sniper saw a trooper swoop down and lifting a Turk into his saddle gallop away; They were all away, they had even taken their wounded. The panting, amazed Turks stood wildly firing after these flying phantoms of the night.

Kara Ismet stood panting his relief when an Australian shout came from the darkness ahead. "Squadron! Sections about! -Action-Front!"

He was still panting when the black crest immediately ahead burst into ribbons of flame, machine-gun bullets thudded around him. The cursed Anzacs were fighting again - would they never know they were beaten?

All through the night the desert battle roared on, the dismounted horsemen steadily keeping up a deadly fire, again and again meeting the Turks with the bayonet when again and again the charge came home. Furiously attacked by overwhelming numbers the Anzac line could here and there be bent, but never broken. Impossible things happened, impossible things made possible only by the night and the fact that the Turk at close quarters becomes madly excited and fires without aiming.

Towards daylight the fate of Romani swung like a slow-moving pendulum, the balance of victory as light as the breath of God or of Allah.

But to Kara Ismet and all the Turkish snipers this battle was particularly terrible. Their special job was to shoot the enemy officers and the machine-gunners. But these were phantom men; it was a phantom battle. And now they were parched with thirst, with throats and nostrils choked with fine sand, their weary limbs trembling.

At daylight the artillery roared into action, all night long there had been hand to hand fighting; neither side daring to use artillery for fear of smashing their own men. But now the howitzers roared, field guns lashed the desert with shrapnel and high explosive. Taubes came diving low dropping bombs and ripping the earth with machine-gun bullets. Against Mount Wellington the Turks surged again and again, their roars of "Allah! Allah!" rolling along the battle-line.

All day long the battle roared and swayed as a tree-branch will bend to and fro when lashed by a conflicting wind. Each side was determined to win. But for Kara Ismet it meant the blessing of at least lying in a sniper's possy although sun and fire made his rifle-barrel too hot to hold. And then came sheer misery, with bitter chagrin.

A horseman, an officer-by Allah a general! appeared galloping over the sandhills directly in front. Reining his horse to its haunches he pointed towards the Turks, his arm sweeping the horizon as he shouted orders to hidden men.

Kara Ismet fired-and missed!

He stared open-mouthed: he had missed an enemy officer and horse outlined against the skyline. His officer looked over his shoulder and screamed at him - Kara hurriedly fired and missed again. His officer reviled him to all the men. The enemy officer appeared again and again, all through the day he appeared to Turkish troops and snipers for miles, galloping, galloping, galloping, always in clear view pulling his horse to its haunches as again and again he cheered regiment after regiment in full view of the Turks. And what a prize! He was a general. The Turkish officers nearly went mad when their snipers, their infantry, their machine-gunners could not hit him.

"Galloping Jack" Royston was not hit throughout the day. He tired out fourteen horses.

By mid-afternoon the snipers fired at him sullenly, firmly convinced that it was impossible to kill him; he was protected by the invisible cloak of Allah.

Hungry, tired, bitterly hurt in his failure to hit the general, near crazed with thirst, Kara Ismet thought longingly of the sweet water wells under the palms of Katia a few miles in the rear. At long last the sun went down and the stars came and the sand-ridges blazed again with stabs of flame. Night was hell with the wild shouts of clashing men, roar of rifle-and artillery-fire, nerves shrieking as bloodshot eyes stared waiting some mad counter charge by the phantom horsemen.

At the birth of a tortured dawn an officer kicked Kara Ismet in the ribs. The sniper wearily rose and joined the three other snipers hunched beside the officer. They walked a few yards out in front. The officer waved his arm.

"Dig in," he muttered hoarsely, "they will charge soon. When they do, shoot their officers!" He showed his teeth at them, then disappeared into the gloom.

Numbly the snipers separated. In front, but a little to the right of them, was a tiny oasis sheltering a rearguard. Behind the snipers their infantry would mass. In this little portion of the line at least the snipers could never retire.

Searching quickly before the coming dawn Kara Ismet picked a bush around which the sand was very soft. In front of the bush was clear ground. Feverishly he dug a trench, lay in it, then covered all his body that he could with sand. He poked a little space through the bush for his rifle, then waited for the dawn.

They came. Leaping from the earth, a long line of men with dawn-light tipping their steel. Gaunt, haggard, smeared with grime and sweat, bloodshot eyes terrible as they gathered speed and with a mad yell were into the first oasis full of Turks. Shouts, screams, curses, thuds of blows, mad, swaying confusion while the Turks went down and down and down.

Then the madmen burst through with an officer and three others leading them. The snipers sighted their rifles, four sharp shots. Colonel Onslow went down as the men beside him went down, but with a terrible shout the others leaped over them. Kara Ismet closed his eyes to those terrible eyes, those bayonets; his head slumped to the sand, his outspread hands twitched slightly. He heard screams and thuds then they were over, and there came the mad crash as they struck the infantry behind. It was soon over, the thousand Turks not killed, surrendered. Far along the line waves of hoarse shouting told that similar scene after scene was taking place.

Kara Ismet lay for a long time, his ears hearing the firing, the vanishing shouts, ears terribly strained for sound of a footfall beside him. But none came, the wave of battle had swept over him. At last, inch by inch, he raised his head. And shuddered. His fellow snipers were dead. Many others were dead.

He was thirst-mad. He staggered back over the body of his officer towards Katia, towards the wells. Distantly around him mounted men were now charging while his comrades were falling in far flung, scattered groups, fighting their way back towards water. The ground was littered with equipment and wounded men crying for water amidst the roar of the Taubes, the crash of shells, the recurring thunder of hooves. The thirst-crazed sniper reached the great palms of the oasis, with maddened others he rushed forward to the wells, a struggling mass of blood-stained, thirst-crazed men. He fought for and won his drink, he snatched a handful of green dates and staggered to a big desert bush. Dazedly he began to munch the dates but his eyes closed, and he slumped to the sand in exhausted sleep.

During early afternoon an officer kicked him. He staggered to his feet, dazedly hearing the roar of guns, the crackle of machine-gun fire. But around him all was quiet, hundreds of weary men lying among the palms. Sullenly they rose up and followed their officers to the edge of the oasis. There among the palms they spread out, selecting their possies, tearing down branches to camouflage machine-gun nests.

Twenty snipers stood by. An officer pointed to the front. Wordlessly the snipers walked out into the open desert, they spread out and each man began seeking where to make his fox-hole. Weary and dazed, they did not feel any pressing personal concern, for their infantry were concealed amongst the forest of palms but two hundred yards behind them.

Kara Ismet craftily chose his possy, dug it out and camouflaged it and himself with bushes. Then he lay down, his rifle resting beside him. He glanced to his front. Out there stretched a gentle slope of hard sand running for half a mile to end in abrupt sand-hills. But there appeared no target although the roar of battle was raging away to right and left. He wondered what was happening.

He guessed that the Turkish Army must have lost half its men, and that the remainder must be in imminent danger of annihilation. The officers would fight a desperate rearguard action and retreat, retreat, retreat.

He shuddered. Oh Allah, Egypt! Would they now have to retreat over that desert pursued by these terrible horsemen-infantry-horsemen. . . . What were they really: horsemen or cavalry or infantry? They were devils such as no Turkish Army had ever fought before. His head dropped on his arms-asleep.

He awoke with his heart in his throat; bullets had whipped up the sand not a yard from his rifle-muzzle. With thumping heart he realized it was a burst from a German machine-gun.

Just a gentle hint to the snipers to keep awake.

Billy Sing (left) with officers of the 5th Light Horse.

CHAPTER VIII
The Charge

SUDDENLY, in mid-afternoon, Kara Ismet needed no German machine-gun to keep him awake for a line of felt hats appeared over the sand-ridges. Then horses came and a squadron of Light Horse were riding down the sandhill. Grimly, Kara Ismet fingered his rifle, what targets these fools would make. In eager anticipation he snuggled down comfortably to kill. Surely these Anzacs had become fools to expose themselves thus, their victory had gone to their heads. Then Kara Ismet stared, all the snipers stared, for yet another line of hats, of horses, of mounted men appeared over the sandhill, riding steadily down behind the first squadron. By Allah! their lines were tipped with steel!

Among the palms immediately behind the snipers the Turkish infantry crouched low. What targets they would soon have. This would be revenge indeed. These fool horsemen must imagine the oasis abandoned. What a trap they would ride into. The machine-gunners saw to their cartridge belts; soon they would be pumping lead into a thrilling target.

But the snipers were staring, for another squadron appeared. A regiment! With fixed bayonets and on horseback! Kara Ismet felt his hair slowly rising. So different this to the usual action of these wild horsemen. This regiment had come riding out from cover upon a gently sloping plain. And sunlight flashed on myriad points of steel. Surely, if they believed the oasis empty they would not ride with bayonets fixed!

Thus appeared on that brilliant afternoon the 5th Light Horse Regiment, breaking into a steady trot. And now up over the sand-ridge appeared the New Zealanders, riding behind them. The regiments came on in perfect line at the steady trot.

Then guns roared away behind Kara Ismet, shells whined overhead and burst above, but behind the advancing men. Kara Ismet shivered, for the gunners had not got the range. The squadrons came steadily on. Now they broke into a canter, a wonderful sight. Rank upon rank of horses with shrapnel burst like pretty clouds exploding above them.

Sudden stuttering of machine-guns, crackle of rifle-fire, whining of shells. The horsemen broke into a gallop as a long, wild shout sped down to the oasis. There came a rising thunder of hooves, madmen were standing in their stirrups and their hoarse roar chilled the blood in Kara Ismet's veins. They were almost upon him, upon him! He stared at the heads of the eager horses, horses as mad as the men. Seizing his rifle he leaped up and ran, ran as he had never run before. And the others ran with him, caring not that they ran straight into the fire of their own infantry.

But these too, were now running, running back into the palms. Their terrible hardships, the desperate fighting they had gone through unnerved them. They were unable to withstand the thundering squadrons and above all - the steel.

I will never forget that mad, mad charge. And when close to the oasis the sudden breathless fear, for by standing in the Stirrups we spied some of the foxholes, expected the rattle of fire and machine-guns with the crash of horses but none came, and in moments more we were crashing far into the oasis. The colonel's horse was down.

Only a big morass pulled us up. Dragging our floundering horses from the mud, with machine-gun bullets ripping down the dates in showers, we battled back in among the palms, then on foot set out after the Turk. He was badly rattled. Soon, however, he began to fight; he is a great fighter and perhaps never better than in adversity. We rushed forward shooting at them as they shot back at us; they were firing while jumping back behind the bushes and sand-mounds and palm trunks. We shot through the bushes as we plunged on, while bullets spat through the leaves at us; then as we ran among the sand-mounds we would glimpse gasping Turkish faces as they disappeared around the next bushes, hear their grunts as they leaped back as we rushed their cover with stabbing bayonets.

Around one bush was a crouching Turk, his face frantic with its broken nose and livid scar across its left cheek. I yelled at his rifle-muzzle and the bullet whizzed past my ear as he plunged back into the bushes with my bayonet an inch from him. He got away. But other snipers didn't. A crowd of New Zealanders beside us yelled terribly and we instinctively knew that they were charging a snipers' nest. A white rag suddenly waved from the bushes and twenty snipers knelt there with upraised hands dumbly imploring back the New Zealand bayonets.

We fought on and rushed the wells, those wonderful wells full of the water of God. Around them were Turkish footprints, haversacks and bullet perforated water-bottles. How both sides fought in the Desert Campaign for the wells!

Kara Ismet and the Turkish infantry fighting as even they had never fought before were pressed back yard by yard, running only when the steel was within yards of them. Wonderful fighters the Turks - but they simply could not stand the last final test of the steel.

Kara Ismet was fighting as the infantry beside him were fighting, with every nerve strained, every sense alert, every man crouching for the instant leap back with rifle upraised to ward off the steel. His eyes glaring at the bush shielding him, at the big clumpy bushes to right and left, at the sand-mounds amongst the palm trunks, his rifle levelled awaiting the rise of a felt hat. Again and again he fired at the sway of a bush, again and again caught sight of a brown-armed man running low from bush to bush, from mound to mound. But it was almost the merest glimpse, as again and again he and his mates leaped back to crouch and fire again, with bullets whipping the bushes at their faces, ploughing the sands, whistling past their ears. Then would come a long, bare patch of sand behind them. Desperately, their chests panting, lips thirst-maddened, they would glance at that sand and grouping together in an irregular, crouching line hold the bushes until the Australians were almost upon them. Then run for it in desperate leaps across that bare patch while time seemed eternity and their flying legs were lead. Sobbing, they would throw themselves down behind the bushes and face the Anzacs to give them hell in turn when they too must charge across that sand.

For the second time during that fateful afternoon I caught a glimpse of the broken-nosed sniper. His hot rifle-bolt had jammed, and he was kneeling with the snarl of a maddened animal upon his face as he tried to ram another cartridge into the breech. He sprang up at me with a yell; our rifles clashed, and he leaped back and vanished among the bushes. It was an act such as men automatically do when wound up under the greatest tension, the one wild instinctive leap in which lies the only chance of safety.

Towards sundown we had driven them back a mile. Then their firing began to grow stronger; reinforcements reached them. Officers were hurrying amongst them, urging them to hold on, to hold on. If the Anzacs broke through it would mean the end of the remnant of their army. And they held on.

Anyhow, sundown finished us. Outnumbered, our regiments were scattered far and wide throughout that mighty oasis system. For three nights hardly a man had slept; we had been fighting on a bottleful of water a day; some of the horses had worked incessantly sixty hours without a drink. Throughout the oasis spread the word for every regiment to retire fighting, then mount and ride back to Romani and water.

And then the Turkish reinforcements pressed us hard. The Turkish officers hoarsely spread word that they must take back the Katia wells; that if they failed then every man must perish of thirst on the morrow.

Thirst! Kara Ismet's parched lips burned hot, he felt he could not advance a yard farther. . . . But water! water!

The whole Turkish line crept stealthily forward.

Kara Ismet could not believe his senses, the breathless soldiery glanced at one another, the rain of expected bullets did not come. They pressed forward as men reprieved.

Then the bullets came. It was a desperate fight that.

Now it was the Turks pushing back the Australians and New Zealanders and Yeomanry. The Turks advanced against a cool, steady fire, the men before them gave way only when their comrades were back under cover and firing at the Turks, only then did the others leap back to cover.

Again and again and again the Turkish line was held up, methodically and coolly. It was impossible to break through; they were able to advance only just as fast as the Australians allowed them - no faster. But at sundown Kara Ismet grew madly excited; he thought they had them; they would bayonet every man of them with the dying of the sun. The edge of the palms was so very near; the whole Turkish line was fiercely advancing. Then came a sudden crackling, the thunder of hooves galloping away over the fallen palm branches. The Turkish line in mad joy pressed forward to be met by a terrible fire. They dropped down behind the bushes, long ribbons of flame spurting from their machine-guns in the fast-gathering gloom. Another thunderous crackle, hooves again dashing out on the desert. Another squadron had got clear away.

Mad at the escape of their prey the Turks rushed forward; but coolly shooting men and a well-placed machine-gun dropped numbers of them into the bushes. Running, crouching low among the mounds, they pressed forward again. Kara Ismet saw a black cloud among the palms, he took no notice of it, being too busy firing at the flame which spat at him from ahead. Then he saw shadowy forms leaping up on the black cloud which dissolved into a mass of horses. Kara Ismet fired as he had never fired before, his mates rose with a hoarse yell and came racing on, but the cloud was moving away with here and there outflung arms to help some gasping straggler on to a horse. There was a furious crackling, then the squadron thundered out on to the hard sand outside the oasis, and away. The last squadron had gone, taking themselves and even their wounded into the night.

The panting Turks stared at one another, Kara Ismet stared at the mingled rage and relief quivering on the faces of his comrades. Then, as if at one great unspoken urge they called on their last ounce of physical energy for one terrible rush to the wells.

Miles away, in the great oasis system, the mounted men were similarly retiring.

Men of the 6th Light Horse taking a break from the campaign,
photo by Barney Haydon.

CHAPTER IX
Live to Fight Another Day

KARA ISMET was kicked and shaken and kicked again. Dazedly he grew aware of the shadows lying around him being similarly awakened.

"Quick! The Turkish Army must retire. Now! The Anzacs will return with the dawn. We retire on Oghratina where reinforcements and rations await us. Quick! Every man awake! Awake!"

Dumbly the exhausted men stumbled back among the palms. To Kara Ismet it was a nightmare. of weariness, of darkness and shadows and starlight all hazily moving, murmuring underfoot.

Presently his wits began to function. Oghratina, yes. It was at Oghratina that they had toiled weeks ago, digging the redoubts. The officers would be glad they had dug those redoubts. Oghratina! Yes, he remembered. It was at Oghratina where long ago they had slaughtered the Yeomanry. There was the mist; they were creeping over the desert; dim laughter came out of the mist, then the crash of rifle-fire, the flame from the machine-guns, the screams of horses. That was the great raid. As they stumbled on among the tussocks Kara Ismet went over it all again. And now it was their turn. The desert always won. How many men had he seen die in the desert throughout the years he had been a soldier, a sniper? Armies always marching, always stumbling back again. Thirst, exhaustion, death. The desert always won.

Urged by the desperate officers the Turkish forces stumbled back, ever back towards the safety of the redoubts. While miles behind them the victorious army stumbled back also, ever back towards Romani and water. Black clouds upon the desert were regiments of exhausted horses, and pillowed upon their bellies and asleep among their legs were the exhausted men. To a steady command such a cloud would uneasily ripple and move and bulge, then the horses were on their feet, the men wearily climbing into the saddles or else stumbling on, leading their horses. It was the desert - the desert always won.

Thus had the armies of the Egyptians, of Antony and Cleopatra, the Roman centurions, and Napoleon, and the Arab and the Turk marched and fought and lost and won. They stumbled on and on for the desert always won - always won.

Next morning found Kara Ismet in a fox-hole, hungrily eating rations hurriedly issued to the troops. He knew that scattered far to either side of him were other hungry snipers while close behind them was a strong, picked rearguard of Turkish infantry well supported by artillery. The remainder of the army was desperately struggling back to the wells at Bir el Abd, some twenty miles away. The rearguard knew they must hold the enemy to the last man.

Kara Ismet suddenly ceased eating as from an oasis he saw emerging several scouts of the dreaded Felt Hats. He aimed steadily, but before he fired came a "crack!" and the horse of a scout plunged and fell. The others galloped into cover, the dismounted horseman running forward with a bullet through his hat from Kara Ismet's rifle. Then a scattered squadron came galloping from the oasis and under the sniper's bullets galloped into the bushes and on until they disappeared into a shallow depression just deep enough to hide their horses. Not a man had been hit although Kara Ismet and his mates had steadily worked their rifle-bolts. These galloping horsemen with their cunning for cover were flying targets that came only to vanish, to appear, to vanish again. The Light Horsemen dismounted and ran behind cover closer to the hidden snipers. An infantry machine-gun began stuttering, a burst of rifle-fire to the right heralded the approach of another squadron.

Kara Ismet peered eagerly along his rifle, searching for a target. Heavy rifle-fire broke out from the Turkish rearguard. Every man of them knew that if the horsemen broke through it must be only because they were annihilated. From Hod el Sagia came distant sound of fighting, guns boomed from Hod Abu Darem. Taubes roared overhead to swoop and drop bombs in the oasis. Kara Ismet listened to the shattering explosions with pleasure. He was glad he was where he was.

But the mounted men only pressed forward until they came in close contact with the sheltered rear-guard. The breathless day grew into a far-spread roar of musketry that crackled away in waves to come again then die away to suddenly burst forth once more.

The Turkish rearguard fought with the despair of men who knew they had nowhere to retreat to. Their enemies consisted of a few brigades. The horses were done at last; the active brigades had only to keep in touch with the enemy.

Kara Ismet soon sensed this. Very relievedly, he settled down to steady shooting. The sniper is a deadly foe when shooting thus.

Next morning the rearguard fight broke out again. Kara Ismet eagerly sought targets throughout that day too. He was much refreshed, for he had lain in his fox-hole for twenty-four hours, and had been fed and watered too.

Kara Ismet saw a new dawn come and with it the mortification of his life - men at his very rifle-muzzle and he could not shoot a man!

With the dawn they came: eight men, two sections. One wide-spread section came galloping among the bushes almost straight towards him, the other galloped away in behind him. He held his fire, waiting with bated breath. Had they found him? Realized that his possy was there?

The first section dismounted only fifty yards away, the four scattered men then walked their horses towards one another. These were sheltered by a natural trough in the sand; he could see the four horses' heads level with butts of the bushes; see their big brown eyes as they patiently stood there held by the horse-holder. The other three Felt Hats sat near him, eagerly peering amongst the bushes towards the line of the Turkish rearguard.

Kara Ismet lay still as a mouse, staring over his rifle sights at a nice, eager-eyed boy, his forehead covered with tousled hair in need of a barber. The sniper's eagle eyes could even see that the boy's eyes were brown, his sun-browned face clear and fresh, a smile upon his lips as he eagerly peered for sign of a Turkish head away back there on the Turkish redoubts. This, an Australian advanced outpost.

Kara Ismet, longing to fire, dared not. He could take life but his would certainly be forfeit. He could kill the boy soldier with the greatest of ease, but the others would instantly disappear, with hell in their hearts. Then would commence the man-hunt. And the section that had disappeared away behind him, they were an outpost too. Seven men, determined to kill, four behind him, three in front, all determined to kill Kara Ismet. He might possibly kill two more but if so he knew that he would never, never, see another sunrise.

Kara Ismet had killed many men, he had fought in the Bulgar campaign, had fought against the Arabs, the Circassians, the Russians, the Kurds, the Greeks, the Anzacs on Gallipoli. Seeing so many men die; he understood full well a man can die but once. And he did not want to die. No, it was foolish. He would live to kill another day.

All the live-long day that outpost stayed there, harrying Kara Ismet's nerves. The fear that any moment he might be discovered, the almost irresistible urge of the man-killer to pull that trigger. Several times in half-laughing, half-cautious tones, his comrades urged the boy to be careful, to keep his head lower. At last the rough-faced section leader swore violently at him. Reluctantly, the boy crouched lower amongst the bushes. Kara Ismet smiled as he sensed the meaning in those angry words:

"Look here, Jimmy, you stupid coot, I don't care a hell's damn if you get a bullet through your own head, but I'll punch your blasted nose for you if you draw fire on the section. Keep your head down, you stupid fool!"

As time wore on, Kara Ismet grew increasingly curious. It was evident that the section had not the slightest idea of his presence. So long as they remained hidden and did not come creeping amongst the bushes, he and they were safe.

Eagerly he peered, noting closely what manner of men these were; he had never seen them at such close quarters before except in the mad excitement of the bayonet charges. A man is only conscious of raving eyes and shouting mouths and stabbing steel at such moments. He certainly had seen a few, a very few of these men dead. But no prisoners. It was easier far to catch a desert deer than an Australian prisoner, even though the officers offered such big sums for even only one. Kara Ismet peered with increasing confidence from his fox-hole. The boy soldier was a nice type of lad, he who swore at him was a rough-faced soldier who now yawned good-humouredly, then thoughtfully spat up and over the bushes as if measuring distance. Then growled.

"Betcher a tin of bully-beef 'gainst a packet of fags I can spit further'n you, Blue."

"Yah!" said Blue and spat out over the bushes. "He's beat you," laughed the boy;

"Yah!" growled the bad-tempered one. "That was only gettin' the range."

He closed his mouth, took a deep breath, puffed his cheeks, then tilting back his head spat mightily. They all stared, Kara too, to see where the spits would fall.

"It's up to you, Blue," declared the boy.

Kara watched in interested amazement as Blue drew himself up to have his spit. Blue was long and thin, with a shock of red hair and red hair on his long, bare arms. He handled his rifle as if was part of him, it was never in his way, it was always where it should be, he handled it as easily and comfortably as he handled his own long arms. Kara Ismet watched his lantern-jawed face with comradely feeling and hoped he would win the spitting. Instinct told him that this man, Blue, was a rifle-shot.

The fourth man, the horse-holder, Kara could not see, only his hat among the bush roots as he squatted there in the depression holding the horses' reins. He was smoking, but so carefully that no trace of smoke showed up amongst the bushes.

Suddenly, whizz-zzz, and the bad-tempered one had flattened the boy to the bushes.

"Told you so," he snarled. "A bloody sniper has spotted you."

The tone of the words froze the blood in Kara's veins. He crouched like a startled mouse in his fox-hole. The three men were now lying flat behind the sand-mounds, their rifles out-thrust, staring towards where the bullet had come. Kara Ismet peered fascinated. Many a time he had sniped at these men, now he was watching them being sniped while they sniped a sniper. In all his experiences of war he had never watched anything like this.

"Got your glasses, Blue?" growled the bad-tempered one. And there was that in his voice that chilled Kara's veins.

"Yairs," answered Blue.

"Well, set your bo-peeps on him while I draw his fire."

Blue leaned on his elbows holding spy-glasses to his eyes. Steadily he searched the sandy, bush-covered hillock away to their left front.

"Can't see any sign of the cow."

"Well, watch out."

The bad-tempered one broke a stick, carefully fitted his hat to it, then gently raised it above the bushes in perfect imitation of a peeping man. Then lowered it in disgust.

"Too old in the horns to be caught with baby food," he growled. "Well, here goes."

Peeping round, he crawled out in front a few yards. Kara watched breathlessly. The bad-tempered one was availing himself of excellent cover, crawling behind the sand-mounds piled about the butts of the bushes. He stopped now at a long narrow mound about three feet high. To expose himself above that mound would mean to expose himself to the distant sniper. But over that mound Kara knew was a deep trough in the sand.

"Ready?" called the bad-tempered one.

"Yeah!" answered Blue.

The bad-tempered one raised his hat just above the bush. No shot came. Calmly he placed his hat on his head, then in an instant was up and over the mound. A bullet came instantly but the bad-tempered one had simply leaned over the mound to sprawl in the trough.

"Spot him?" he yelled.

"Pretty near," called Blue, "to a yard anyway. He held his rifle-muzzle just a bit too low and I just seen a whip of dust."

"Where?" asked the boy eagerly.

"Keep your head down?" growled Blue. "There'll be plenty of time when Joe comes back. If you fire now and spoil the fun he'll kick the backside off you."

Presently, the bad-tempered one appeared down by the horse-holder. He came crawling back.

"Landed on me guts when I jumped," he growled hoarsely. "Knocked the blasted wind out of me. Just wait till I get me sights on that Jacko."

He settled himself very comfortably, his rifle poked

between a bush, resting nicely among the bush butts on top of the mound.

"Now where is the blighter?" he growled.

"Four hundred yards front. See that patch of white sand down the slope of the little hill with green bushes ringing it?"

"Yes."

"See dead centre in the white patch?"

"Yes."

"Run your eye straight up. See three big bushes on sand clumps almost in line?"

"Yes."

"Well he's lying somewhere behind the roots of the middle bush."

"Oh, is he. Well he won't be lyin' there long, not unless he's dead meat. Now, young 'un, got the range?"

"Yes," answered the boy excitedly.

"What's your sight up to."

"Three hundred yards."

"You bloomin' goat! How the hell the recruitin' sergeant ever thought we'd make a soldier out of you I dunno. Didn't you hear Blue say four hundred yards?" .

Shamefacedly the boy raised his sights.

"Now," growled the bad-tempered one, "take your time and take a steady pull. When you've got the butt of the bush sighted, count three, then fire."

Coolly the three levelled their rifles. Kara Ismet stared fascinated, he knew that three bullets from three slightly different angles would soon be ploughing into the butt of one bush.

The three rifles spoke almost at the same second. Then the bad-tempered one and Blue were working their rifle bolts with great rapidity, pumping coolly aimed bullets into the bush. The boy fired slower and more excitedly.

Kara Ismet knew that the two other men were pumping bullets with such quickness and accuracy and at such varying angles into that bush that if a sniper lay there behind he could not live!

Kara shivered. The one. fact that had given him such long life was that he was always so careful, so cunning. He never gave a chance away; he had killed so many men who gave a chance away. He could imagine that sniper out there, the bullets pouring into his fox-hole.

"Fifteen of the best," growled the bad-tempered one cheerily, as he slipped a fresh clip into his magazine. "If he was behind that bush he's either there no longer or else he's dead meat. Now we'd better keep our eyes skinned and watch the Jackos in the redoubts don't play any tricks on the boys in the Hod below. And, young feller me lad," he exclaimed fiercely, "if you raise your nose above those bushes again, I'll knock your block off."

And so the day wore on, a day of thrills for Kara, full of bitter disappointments. Never had he seen such wonderful targets; again and again a full troop of horses emerged from the oasis below, while away across the sandy flat a full squadron was resting, their horses partly visible to him among the palms. But he dare not fire a shot. All day long that outpost lay there right under his nose. And well he knew the hidden outpost behind him was similarly watching the Turkish redoubts.

He grimly promised himself the pleasure of killing at least one of these men at sundown. They would ride away at sundown, he would shoot one as they were riding away. To relieve the strain he listened to the "plip-plop, plip-plop" of the snipers' rifles, the sustained fire of the Turkish rearguard, the roar of the guns; watched the circling Taubes over the mounted men in oasis and sandhill. But always his eyes came back to the bad-tempered section leader, to Blue, and the boy.

Sundown came. Kara Ismet watched the squadrons retiring; admired their galloping as for miles they dodged the bursting shells-their officers always seemed to know where the following shells would be falling. He grinned sarcastically at the thought of the German officers away back with the guns, fuming as they tried to get the range of these unrangeable horsemen.

His fingers were gripping his rifle, almost trembling, fingers that so rarely trembled. But to lie for twelve long hours and actually see the eyes of men and not kill one.

"Right-oh!" growled the bad-tempered one. "Get to your horses."

Blue and the boy crawled down into the sand lane. Kara heard the boy's low laugh as he patted his horse. He did not know whether he was glad or sorry. But the bad-tempered one still lay there, staring through the bushes at the Turkish redoubts. All was growing hazy now, sunset was vanishing into the darkness across the desert. Now was the time to shoot!

Suddenly, Kara remembered that the other three were now down in the shelter of that deep sand-trough. Should he pull the trigger?

The bad-tempered one began crawling away, Kara's trigger finger trembled. With almost a moan he lowered his rifle. If he fired, the bad-tempered one would be dead - but he would roll down the sand-trough to the feet of his comrades. They would instantly hunt Kara. Not only that, there was the outpost somewhere behind him.

Kara still waited, waited until they should mount their horses. Just as they were galloping away to escape the fire of the redoubt, then -

But they did not gallop away - not then. Instead, the bad-tempered one led them silently along the sand channel. Kara Ismet cursed. That cunning old campaigner would lead them to where the sandbank was high towards the Turkish redoubt but low towards the dim oasis. There they would meet the second outpost, in a flash be mounted and, scattering, be at full gallop.

He hurried forward on his knees; still he might get a shot. Then he remembered the flash - it was dark, the flash would immediately catch their eyes. He could hear the yells of the seven as they wheeled and spurred their horses back to him. They would gallop over him and club and trample him to death.

There came a sudden flurry, a pounding of hooves and they were away, phantom horsemen of the night.

For a long time Kara crouched there. He hated to be beaten; he would have loved to have returned to the redoubt to show a new identity disk to string on the "necklace" he carried around his neck.

" 'Tis the will of Allah," he murmured. "At least I live to fight another day."

CHAPTER X
The Steel

KARA Ismet lay uneasily dreaming in a redoubt at Rafa: "The desert always wins. The desert always wins .... But I am not in the desert," his worried mind was protesting. "I am in Palestine."

No. He was plodding doggedly on in that nightmare march of the Turkish rearguard from Oghratina. He was in the shell-torn hills of Bir el Abd. He could feel his blistered elbows in the burning sand; could smell his hot rifle-bolt; saw again the dying hand of that New Zealander, saw the fingers twitching.

"The desert always wins. The desert always wins." Again in retreat, under desert stars with his feet like lead as the miles dragged by. Comrades doggedly marching to the whispering of the sands: "The desert always wins! The desert always wins!"

Salmana! The roar of guns, the wild horsemen silhouetted on the sandhills to come plunging down into the gullies. Always coming; always coming. Kara Ismet moaned in his sleep as the night-shrouded column ploughed wearily back to Mazar.

"Mazar!" He flung up arms to ward off the plunging bellies of horses, the startling eyes of brown-faced men as they leaped down into the snipers' nests.

He dreamed again of the retirement from Mazar, always retiring, always retreating, the sands whispering. One hundred miles of sand.

El Arish! Then Magdhaba. Magdhaba!

Magdhaba won. Magdhaba lost. Flying from the roaring shouts, from the gleaming bayonet-points. How he ran to the roaring throats as redoubt after redoubt fell. He and a hundred others were all who escaped. All those hundreds and hundreds of others were killed or captured. Kara Ismet shuddered; for the redoubts of Magdhaba had been taken at bayonet-point.

He crouched up glaring, ears tingling, heart hammering. Then sighed heavily. Rafa! Of course, he was in the Rafa redoubts. No wild horsemen could ever storm these. The shrouded forms of his comrades lay all about him, their rifles beside them. A German machine-gun was silhouetted in its firing possy.

Sighing with relief he stood up and stared out over the parapet. He saw far into Shadowland for he was staring down a gentle slope on to a plain quite bare except for infant shoots of barley. Upon this long ridge on which was built the redoubts, one solitary tree was silhouetted against the brilliant sky. Men were to die around that tree before the stars shone again. Perhaps some premonition of this had caused Kara Ismet's nightmare. He was not in the desert; that ended abruptly as a brown line a mile or two away. This was El Magruntein near the Egyptian Police Post at; Rafa, the border of Sinai and Palestine.

He had left that cursed desert forever, the desert over which for a hundred terrible miles the wild horsemen had driven the remnant of a beaten army. But they were safe now, heavily reinforced in deep earthworks impossible of attack by mounted men. And behind them, at Khan Yunus and Beersheba, waited 100,000 men. He smiled as he thought too, of the army at Gaza with the redoubts of Ali Muntar made impregnable by the German engineers. He again sighed his relief, rolled his greatcoat about him, and snuggling down in the trench fell into a sound sleep.

Kara Ismet was awakened by urgent boots, startled shouts, urgent commands of officers. He leaped up amongst hurrying men seizing rifles while shouting comrades came leaping down into the redoubt. In a second he was at the parapet.

"Allah!" His heart almost stopped beating. Distantly there came the wild horsemen, long columns of them galloping in the dawn. And away to their flank appeared long lines of hurrying camelmen.

Kara Ismet's heart thumped at the now familiar sight of a brigade breaking into a gallop, those deadly fighters in the felt hats with the eager eyes and leaping bodies that never knew when they were beaten. But Kara Ismet did not know that these terrible men were now eager schoolboys on a wonderful holiday; laughing, shouting, joking on horses like themselves suddenly frantic with delight. Firm ground under them, tender shoots of barley, scarlet poppies of Palestine, a garden of wildflowers amongst tender young grass. No wonder the poor old horses went frantic; no wonder the troopers laughed and shouted. as the thundering hooves kissed firm ground gay with flowers. The desert was behind them. Finish the desert!

The mounted men, New Zealanders, Australians, Yeomanry, the Royal Horse Artillery, the Hong Kong and Singapore batteries, cared nothing that the rules of warfare declared the position before them, impregnable to mounted men. The long, grim ridge with its strong, heavily defended. redoubts looked down upon a plain destitute of cover. The mounted men laughed that they had taken Magdhaba with its redoubts and artillery, its machine-guns and riflemen and Taubes. To hell with Rafa! Here was firm ground underfoot, grass and flowers. The desert was finished.

Rafa was a classic fight; won like Magdhaba, at sundown with only minutes to spare between victory and defeat.

Kara Ismet fought all day with a tense coolness nearly breaking under anxiety towards sundown. Galloping men had early cut off the redoubts from chance of relief, from Khan Yunus and Shellal. And now each redoubt was encircled by men. This was no fight in which the snipers could do their particular work then slip over a ridge and away, or vanish back to the shelter of the rearguard. Here, every man, every battalion in every redoubt was surrounded. It meant a fight to a finish. He fired in desperation while as sundown drew near the heaviness of his dream was upon him.

"Fight on! Fight on!" urged the officers. "Shoot to kill! Hold them off until sundown and they must retire for water. Shoot to kill! To kill! Remember Magdhaba!" : As if he could ever forget Magdhaba!

All through the day those irregular lines fighting in their own peculiar way as infantry kept drawing nearer and nearer, only a few yards at a rush, coming ever nearer, nearer. Inconceivable that a man of them could live under the hailstorm of bullets ripping the far-flung battlefield. Unbelievable that far more were falling in the sheltered redoubts than out on the open plain. Upon the redoubts rose a haze of dust as streams of bullets swept the parapets lined with Turkish heads. Kara Ismet's eyes were half-blinded with flying gravel, his broken-nosed face smeared with sweat and blood as with one boot wedged against a dead comrade he tried desperately to shoot straight. Long lines of men with even gaps in their lines continually leaped from the plain and rushed forward to throw themselves down, while feverishly the 'Turks altered their sights to get the ever varying range. The firing would again grow into a roar as streams of bullets came whipping the parapets, then suddenly those gaps leaped up as men and they too dashed. forward to throw themselves down with their fellows. All day long coming nearer and nearer; those that ran forward protected by the fire of those lying behind. Then those who had been lying behind would be protected by those who had run forward. How could the Turks pick off these irregularly moving men while concentrated fire skimmed the parapets and the range was constantly altering.

"Their officers, you dog, their officers!" Kara turned with a snarl to the fierce grip on his shoulder, but the officer's face turned bloody and he slid down.

"Dog!" gasped Kara. "You got what I would have got had you not struck me!"

He turned to his firing again and his heart was thumping painfully.

With the greatest bravery the Turks, again and again, stood shoulder high in their parapets to get a better view of their enemies as these rose in their short rushes, but at such times they lost fearfully. At such close range the British and Turkish batteries worked their guns to a ceaseless roar, until the mechanisms became too hot to touch.

And here was a strange sight! Under the British shells

screaming towards the Turkish redoubts and the Turkish shells screaming towards the British guns Bedouins were working in the young barley-fields. For centuries these nomad desert tribes had seen countless armies come and go; witnessed countless battles. But when darkness came these jackals of the desert would sweep like shadows over the battlefields to cut the throats of the, wounded, and steal what they could of the wreckage of war. Now, men and women toiled with wooden ploughs while over the Turkish redoubts planes, the modern terror of the air, swooped and dived and fought. In fierce dogfights British and Australian planes drove off the Taubes, and were now raining down bombs on to the redoubts and artillery. By late afternoon the far-flung rings of men were drawing closer and closer around each redoubt.

Excitedly, Turkish officers ran along the trenches telling the men that thousands of reinforcements were hurrying towards them from Khan Yunus and Shellal.

"Hang on!" they cried. "Fight on! Remember Magdhaba! "

But Kara Ismet's heart froze, for he realized the wild horsemen must know the news also. He saw the New Zealanders rise and in a great circle come running towards the redoubt. They were down again, the redoubt was struck by a sheet of flame as he rammed fresh cartridges into his rifle. As smoke and dust cleared he saw the men coming again and he fired and fired and fired, hearing above the roar of battle the shouts of an officer.

"Fight on! Fight on! Remember Magdhaba!"

As swiftly he reloaded he just saw the Camel Brigade rise and the ring of men dash forward towards the largest redoubt. As he reloaded again the earth out in front leaped to a ring of men and the setting sun gleamed on steel. His heart rose to his throat. Could they hold them back?

Frantically the Turks fired, standing breast-high, their legs wedged between the dead and dying. Under the terrific fire their redoubt was actually "smoking like a volcano" - the very words used in the official report.

To a shouted order they fixed bayonets and Kara Ismet, now like a mad, blood-spattered beast, glared down at the men racing up the redoubt. At the sight of Turkish steel their whole circle roared as they leaped to the challenge, and Kara Ismet glared at their officers and men racing one another to the charge. Then rose to burst above him a wave of flying bodies, blazing eyes, screaming mouths as the steel plunged down. He lunged up with a scream.

Stiff luck, Kara Ismet! Could he only have lasted moments longer! To a ringing clash of steel men went down, then suddenly the Turks dropped their rifles, knelt with upspread arms. Instantly the New Zealanders stood, statues in the maddest excitement that can come to humankind, their bayonets at the throat of the Turk. Then they laughed, a roar of laughter from heaving chests. They slung their rifles and panting, beckoned that they were friends.

From the great redoubt the roaring Camel Corps now were charging the Turks. Firing wildly, as always when under terrific excitement, the Turkish fire sped high. The camelmen were leaping down amongst them. Then up went white flags; the Turks were down on their knees.

The mad roar turned to laughter; the Australians were shaking the hands of the dazed survivors. How many others could do that? Fight all day against a heavily entrenched foe that fought right to the very bayonet-point, then laugh and spare him in the height of maddened excitement.

All over Rafa they were charging redoubts under the setting sun. Redoubt after redoubt fell. Rafa was won.

Turkish guns taken by the Light Horse at Rafa.
Photograph by Barney Haydon.

CHAPTER XI
The Diary

"AW!" growled the bad-tempered section leader. "There's all the souvenirs of a Turkish Army here, and you must choose that!"

The boy soldier held a Turkish diary in his hand and was looking down half curiously, half pityingly at what had been a broken-nosed Turkish sniper.

"Why don't you pick up an Iron Cross from a German officer?" suggested the section leader. "Or a crescent buckle from a Jacko. There's plenty of them about."

But the boy shamefacedly put the diary in his haversack.

"This sort of seems more personal," he said lamely.

"Poor beggar. His life story might be in it."

"His life's gone now," smiled Blue. "And you can't read it. It's in Arabic."

"Pick on something sensible," growled the section leader, "if you want a personal souvenir of Rafa. Here, this is something more like it."

He was bending, untwisting something from around the dead sniper's neck. Curiously he held up the necklace. It was a leather cord strung with metal identity disks.

"Twenty-seven!" counted the section leader.

"That's twenty-seven men he's killed for certain."

"A good shot!" said Blue professionally. "I wonder how many he killed that he couldn't get the identity disk off?"

"Mostly Yeomanry," said the section leader as he examined the disks; "mostly 5th Brigade. He must have been at Oghratina. Here lad, take it. This is a proper souvenir."

But the lad hung back. "I'd rather remember him by his diary," he said.

"What do you want to remember him for?" demanded the section leader. "You've never seen him before, and he's never seen you. If he had he would have given you a souvenir for sure - a bullet. But come along, the boys are moving, it's nearly dark. We'd better scout out and pick up the wounded before the Bedouins cut their throats."

It was some days later, under the palms of El Arish. Some of the boys from my old regiment were curiously examining the victors' souvenirs, the diary among them. It was neatly written in Arabic.

"Get the brigade interpreter to read it out for us," suggested Blue. "He reads Arabic like a baby drinks mother's milk."

And thus we listened to the story of Kara Ismet. From his adventure. of the broken nose to the Russian bayonet and his hatred and dread of the steel ever since. His adventures in other campaigns. Then from Gallipoli to Romani. Then Katia. Then Oghratina, then-

"Christ!" exclaimed Blue.

For Kara Ismet had minutely written of his ineffectual day waiting to shoot the outpost that was the bad-tempered section leader, Blue, the boy, and the horse-holder.

"If that don't beat a book!" exclaimed the section leader.

"That's our outpost he's writing of."

"It is," said Blue grimly. "Go on," he nodded to the interpreter.

"'But for the will of Allah,' " read the interpreter, " 'I would have put a bullet right into the ugly face of the bad-tempered one.' "

"If only I had him here!" growled the section leader, "I'd punch his broken nose so his own mother wouldn't know him."

"'I could have shot the boy too,'" the interpreter went on, "'and added his disk to the ugly one's. But the one I liked best and the disk I would have valued most was the red-head they called Blue. He was a man. He could shoot. But Allah willed it not to be.'"

"He doesn't seem to have taken any account of me at all," said the horse-holder in injured tones. "I was of no account."

"If it don't beat the band!" said the bad-tempered one in awed tones. "Just think of that snake lying there all the time."

"We were jolly lucky," grinned Blue.

The diary went on and on, right to the last few hurriedly scribbled lines.

"And my dream spoke to me: 'The desert always wins- the desert always wins!' And now they have come. Ah! 'tis the will of Allah."

Thus the diary ended. The boy put it thoughtfully away. Later he threw it on the sands. Blue looked up inquiringly.

"I don't want it now," said the boy. "I don't want it; I don't know why."

It was left there when the troops rode away. I've often wished since I'd kept that diary. The full story of Kara Ismet would have made a grim, thrilling book.

I never dreamed of book writing then and besides I had enough trouble keeping up a diary of my own.

Time went on. The dead were forgotten in the long, piled-up trenches at Rafa. And a greatly reinforced army rode into Palestine, towards the frowning heights of Gaza.

ION 'Jack' IDRIESS was born in 1891 and served in the 5th Light Horse in the First World War. He returned to Australia to write *The Desert Column*, which was published following his huge success with *Prospecting for Gold*. He went on to write 56 books and was largely responsible for popularising Australian writing at a time when local publishing was still not considered viable. A small wiry mild-mannered man, Idriess was a wanderer and adventurer, with a vast pride in Australia, past, present and future.

ETT IMPRINT has published new editions of these books:

Prospecting for Gold (1931)
Lasseter's Last Ride (1931)
The Desert Column (1932)
Flynn of the Inland (1932)
Gold Dust and Ashes (1933)
Drums of Mer (1933)
The Yellow Joss (1934)
The Cattle King (audio) (1936)
Forty Fathoms Deep (1937)
Madman's Island (1938)
Headhunters of the Coral Sea (1940)
Lightning Ridge (1940)
Nemarluk (1941)
Sniping (1942)
Shoot to Kill (1942)
Guerrilla Tactics (1942)
Horrie the Wog Dog (1945)
The Wild White Man of Badu (1950)
The Red Chief (1953)
Ion Idriess: The Last Interview (2020)

ION IDRIESS

The Last Interview

TIM BOWDEN

Ion "Jack" Idriess (1889 – 1979) is recognised as one of Australia's great storytellers, having published over 50 books including the Outback tales of *Lasseter's Last Ride*, *Flynn of the Inland*, and *The Cattle King* alongside major works on the histories of Broken Hill, Broome and Cooktown.

This book is his last interview in 1975, prompted by the then young Tim Bowden, for a possible ABC Radio program that did not eventuate. With renewed interest in Idriess and his life, within this book Idriess talks of his early years in Broken Hill, he tells of his earliest writing for the *Bulletin*, on living and photographing Aboriginal tribes in the Kimberleys and Cape York; on the writing of his books like *Madman's Island* and *My Mate Dick*; his life with the pearlers of Broome and Thursday Island; on the joys of prospecting, living in the Wild, on Lasseter and his diary. Full of colourful characters and true stories, Ion Idriess allows us into his unbridled enthusiasm for Australian and Aboriginal history.

LIMITED EDITION OF 100 COPIES, 124 pages, illustrated with Idriess timeline, numbered, in colour

Paperback edition, black and white photographs throughout, 124 pages, illustrated; for more information write to ettimprint@hotmail.com